# Common Sense COVID-19 Guidance for Everyone

## A Quick Reference & More



Dr. Etch Shaheen, MD

Published by Knowledge is Power Publishing, part of On Call Enterprises, Inc. P.O. Box 4238, Lake Charles, Louisiana 70606. To contact the author, address letters to: Dr. Etch c/o KIP Publishing, P.O. Box 4238, Lake Charles, LA 70606.



**The information contained in this booklet is only for informational purposes and not meant to serve as recommendations, guidelines or advice. It is subject to change as more information and knowledge is learned about SARS-CoV-2. If you have any symptoms or questions of any kind related to SARS-CoV-2 or your health, you should immediately contact your doctor or a qualified medical expert. Any information in this booklet is offered as a resource and not intended to imply endorsement nor can its accuracy be guaranteed. No one should make any decision regarding their health, behavior, public health decisions, medical treatment, medications, or any other aspect based on what is contained in this book. There is no guarantee that you or others will not become infected with SARSCoV-2. It is recommended that you follow the recommendations and advice of the Centers for Disease Control and Prevention (CDC), your public health officials, and your physician for the most current and up to date recommendations, guidelines, and requirements. The author and publisher assume no responsibility for anyone's healthcare or other matters, including but not limited to decisions related to whether to seek medical attention, what type of precautions to take, where to seek medical attention, what provider or entity to choose, what medical related decisions to make, or any other decisions or actions, etc. Such decisions are important, and you should consult with trusted and appropriate professionals i.e. your physician, public health officials, CDC, etc., who are experts and knowledgeable about the particulars of your specific circumstances.**

ISBN 978-1-952827-22-8

# Common Sense COVID-19 Guidance for Everyone

## Introduction

---

While working on the next book in the COVID-19 Series: *COVID19: Consequences & Perspective*, I was contacted by a reader of my first COVID-19 book called *COVID-19 Lockdown: Unreported Truths & Perspective.* The reader suggested that a simple reference booklet would help many in the public. She reported that my book is full of lots of valuable information but suggested that a simple, short reference would be helpful to those who don't want to read a lot but prefer short simple answers to common questions that many may have or that would easily understand why the precautions are necessary. Being someone interested in helping others, respecting my readers' input, and reflecting on the matter, I decided to follow the suggestion of my readers. This booklet will provide some guidance based on what experts have said and answer some questions that many may have that can help you understand why experts recommend or suggest the things they do. There is a simple Quick Reference that one can look at and immediately get the basic guidance on what we should do, followed by questions and answers that provide more insight, background, and important bits of information.

There will always be some risk to becoming infected with SARSCoV-2, the virus that causes COVID-19, just as there are risks and benefits to everything we do and do not do. This is a fact that many seem to ignore or not realize. There are certain things that we can all do, or avoid, that will lower the risk of us becoming infected with SARS-CoV-2 and developing COVID-19. Any single item can help reduce the risk but the more items you combine, the more you can lower the risk. The more we do, the more we can lower the risk to ourselves and to others.

Please keep in mind that one cannot possibly cover every possible scenario or know the exact circumstances that apply to you individually i.e. state of health, underlying medical conditions, risk factors, risk tolerance, importance of various activities to you, etc. Any guidance found in this booklet is with the understanding that we should:

1) **Always use good judgment**. There is no substitute for good judgment based on accurate information and applying it to your specific circumstances.
2) **Use common sense**. Things that make sense usually do so for a reason. Do not ignore what makes sense. It may not always be correct but many times it is.
3) **Inform ourselves**. Arm yourself with accurate information. Refer to the CDC website and public health officials and stay current. Recommendations, guidelines, and guidance change as we learn more. Just be sure that the source of the information is reputable and reliable.
4) **Base decisions on the risks and benefits** of those decisions. Take into consideration the information that we know, the risks and benefits of any options we have, or decision that we are considering, and use good judgment and common sense to choose what is best for our particular situation.
5) **Risk.** With few, if any, exceptions, there are always risks and benefits to any decision or action that we take. This is a part of life and we must accept and acknowledge

> this. We all take risks everyday often without thinking about it. I.e. taking a medication, operating a motor vehicle, walking across the street, going hiking, etc. Unfortunately, there are times when we face certain challenges when there may be no "perfect" option, or a situation where no one is harmed or inconvenienced i.e. COVID-19 pandemic. Sometimes no matter what we choose, there are bad outcomes. In such situations, we must choose an option that may not be perfect but is the least harmful or is the best realistic option available.

There are certain things that we can do which in most cases should have greater benefits to us than the harm they create. As a convenience, a quick reference is provided for you. I begin with a convenient Quick Reference for those who want a short and simple version. For those who are curious or want more information and details about the virus, why we use 6 feet as the standard social distancing space, masks, hygiene and other matters related to SARS-CoV-2, the Coronavirus that causes COVID-19. Thank you for purchasing my booklet and doing your part to help keep yourself and others safe and healthy.

# Ways to Stop the Spread of COVID-19 A Quick Reference

---

- ## Social Distancing

  **Maintain a MINIMUM of Six (6) feet physical distance between you and others. The greater the distance, the less the risk of spreading the virus.**

- ## Face Covering

  **At a MINIMUM, wear a CLOTH Face mask. Respirators i.e. N100, N99, N97 & N95, without exhalation valves are most protective to the wearer and others followed by surgical masks, multilayered cloth masks then single layer cloth masks. Respirators > Surgical > Multi-Layered > Cloth > Plain Cloth Masks**

- ## Wash Hands Frequently

  **Wash for a MINIMUM of 20-30 seconds with soap and water**
  **ALWAYS wash hands <u>Before</u>:**

- **Eating, touching eyes, nose, or mouth, putting on mask, caring for others**

**ALWAYS wash hands <u>AFTER</u>:**

- **Taking off mask, touching mask or anything that may be contaminated, touching or caring for others I.e. handshake, care provider etc.**

- ## Sanitize Hands When Unable to Wash Hands

**If unable to wash hands, use hand sanitizer containing a MINIMUM 60% Alcohol concentration (i.e. Isopropyl but <u>NOT methanol</u>)**

- ## Disinfect Surfaces Frequently

**Use effective disinfectants that kill viruses i.e. chlorinated bleach mixtures i.e. household bleach, 60% or greater concentrated alcohol, etc. to clean any surfaces that may be contaminated, particularly those that are commonly**

**touched or that may be contaminated such as:**

**Door handles, doorknobs, hand railings, countertops, sink handles, phones, laptops, remote controls, mailboxes, keys, car interior: radio, garage remote control, handles, steering wheel, controls, etc.**

- ## See Your Doctor & Stay Home
  - ➢ **See your doctor, or go to the emergency department (ER), immediately if you are having difficulty breathing or any significant, or concerning, symptom**
  - ➢ **You may, or may not, have COVID-19 but if you are not well, you still get the appropriate diagnosis and care**
  - ➢ **After seeing doctor, follow his/her advice and avoid exposing others by only leaving home if necessary and practicing good hygiene i.e. hand washing and disinfecting surfaces**

*For more details and information, see the Questions & Explanations Section*

# Questions & Explanations

---

## Why do we Socially Distance?

Social distancing, correctly known as physical distancing, is used to help lessen the risk of spreading the virus. It is believed that the most likely way that SARS-CoV-2 is spread is from someone breathing in, or being inoculated by, the virus that is contained in droplets or aerosol form from someone else that exhales, coughs or sneezes it out. I.e. talking, laughing, singing, breathing, coughing, etc. It is believed that the most common way to become infected with SARS-CoV-2, which causes COVID19, is by person to person contact, specifically by breathing in the virus that may be contained in small droplets or in aerosolized form. Droplets are larger particles that can contain the virus such as sputum ("phlegm") spit, mucus ("snot") while aerosols are even microscopic droplets that can linger in the air and can occur when people sneeze or are getting medical treatments i.e. placed on ventilators or getting nebulized breathing treatments.[i] [ii] It makes sense that the farther away someone is from a person who may be infected with SARS-CoV2 or COVID-19, whether asymptomatic or symptomatic, the less likely that person will be to catch it.

## Where Does the Six (6) Foot Distance Come From?

The United States Centers for Disease Control and Prevention

(CDC) recommends a physical distance ("social distancing) of 6 feet between people to lessen the chance of spreading SARSCoV-2.[iii] The World Health Organization (WHO) recommends a physical distance of 3 feet "and only when people are coughing or sneezing."[iv] The reality is no one knows for sure what the exact distance is that will prevent someone from getting SARSCoV-2. Lots of factors can increase or decrease your risk of becoming infected with SARS-CoV-2, the most obvious one is whether you come into contact with the virus, or someone infected with SARS-CoV-2. Since we cannot "look" at someone and know because many are asymptomatic carriers, to be safe and minimize the chance of catching it, we must assume everyone is a carrier.

It is thought that the distance recommendation originated from research related to tuberculosis that was done in the 1930s by a Harvard researcher named William F. Wells. He studied how far droplets traveled that were exhaled by patients.[v] [vi] **He concluded that they traveled about three (3) feet before gravity took over and the droplets would fall to the ground.** From this, **it is believed that the recommendation of 3 feet of space between people with communicable diseases transmitted by droplets, or respiratory spread, began**. This "three-foot rule" for drops has been used since the 1930s and seemed to apply to the spread of viruses that are also spread via droplets including "the flu," SARS-CoV-1 and the Middle East Respiratory Syndrome (MERS).[vii] As mentioned earlier, the WHO still uses 1 meter, or about three feet, as the distance for their social distancing. But other factors can influence how long droplets remain airborne or how far they travel. These include droplet size, air temperature, humidity, air currents, the speed of the droplet i.e. breathed out, sneezed out, coughed out, etc. According to Krys Johnson, a Temple University epidemiologist quoted in livescience.com:

> "Six feet is the average distance that respiratory droplets from a sneeze or cough travel before they settle and are no longer likely to be inhaled by other people. I have seen estimates for social distancing of up to 10 feet if someone sneezes quite hard, does not cover their sneeze or cough."[viii]

**In 2003, during the SARS (SARS-CoV-1) outbreak, researchers found that the virus was transmitted to others on a plane that were as far as six (6) feet away.[ix] It is thought that this may be the reason why the CDC uses six (6) feet instead of three (3) feet of distancing for SARS-CoV-2.** It is reported that the CDC was contacted multiple times to find out the origins of the six (6) foot guidelines but the CDC "failed to comment."[x]

While six (6) feet is the CDC guidelines, it does not mean one cannot catch it if they are 6 feet away. In a study published in the *Journal of the American Medical Association (JAMA)* in March of 2020, Lydia Bourouiba PhD, who studies fluid dynamics at the Massachusetts Institute of Technology (MIT), writes:

> "Protective and source control masks, as well as other protective equipment, should have the ability to repeatedly withstand the kind of high-momentum multiphase turbulent gas cloud that may be ejected during a sneeze or a cough and the exposure from them." And "Peak exhalation speeds can reach up to 33 to 100 feet per second (10-30 m/s), creating a cloud that can span approximately 23 to 27 feet (7-8 m)."[xi]

In plain English, under the right circumstances, it is possible for droplets or aerosol to travel over 25 feet after someone coughs or sneezes. One might argue that 30 feet is a safer distance to socially distance. This is probably true, but one also must weigh the relative risks and benefits and determine what is best for their individual circumstances. Maintaining 30 feet of distance with everyone also has consequences and may not add very much additional protection, particularly if we use other protective measures such as masks, eye protection, good hygiene, disinfecting, quarantining the sick, common sense and good judgment.

Air currents, temperature, and humidity can affect how far droplets travel and travel, as can the size of the droplet itself. In a report on a different study, it was “found that when people walk briskly or run, their bodies create wakes of air that can carry respiratory droplets up to 15 feet.”[xii]

# Are Face Masks Necessary?

The simple answer is YES. As we breathe, cough or sneeze, we exhale or emit small droplets and smaller aerosol material that can contain SARS-CoV-2 and expose others to becoming infected. Since we cannot be sure if we are infected or are carriers (symptomatic or asymptomatic), we must be extra safe and assume we could all have it. Wearing masks helps catch the virus containing droplets, or material i.e. spit, mucus, phlegm, in the mask and lessens the amount of virus spread into the air that others may inhale, or onto surfaces that others may touch. If everyone wears masks, it can significantly reduce the risk of spread.

While there are circumstances when a mask may not be necessary, the fact of the matter is we do not know for certain. Because of this, the saying, “it is better to be safe than sorry” applies. Until we know more and can be more certain, when in doubt, we should always practice common sense and use good judgment. Currently, that includes good hygiene, social distancing, and face masks at a minimum. Otherwise, you expose yourself, your family, and others to possible risk. This is in no means to say that if you wear masks you, and others, will be 100% protected nor does it mean that if you do not wear masks you and others will become infected. What it means it means is by wearing a mask and doing other things, you can lower the risk to you and to others. This may be the reasoning behind many states and localities mandating the public to wear a face mask in public. Considering the potential risks and benefits of wearing a mask, it is a reasonable measure at this time.

# What Face Masks are the Best?

Any mask provides more protection than no mask at all. Under most circumstances, masks include surgical masks and cloth masks. The N95, N97, N99 and N100 “masks” are technically not masks but are respirators. Respirators provide protection to both the wearer of the respirator and those around the wearer. Any respirator that has an exhale valve still provides protection to the wearer but exposes those around the wearer to possible infection from the wearer. **In order of providing most protection to the least are: N100 > N99 > N97 > N95 > surgical masks > multilayered cloth masks > single layer cloth masks > no mask at all.**

Respirators are tight fitting breathing apparatus that will filter a certain percentage of particles larger than 0.3 microns in size. As a reference, a red blood cell is 7 microns in size, or over 20 times bigger than the size that the respirators will filter.[xiii] **A N99 respirator will filter 99% of all particles 0.3 microns or larger while a N95 respirator will filter 95% of the same size particles.**

# How Much Do Masks Protect Us?

A surgical mask is not as good as a respirator but is better than cloth masks or no mask at all. It does provide good protection to those around the wearer and a fair amount of protection to the wearer. Based on a Northwestern University study in updated April 17, 2020, a **N95 respirator filtered out nearly 100% of particles**. Compared to the N95 respirator in the study, here is how other masks performed:[xiv]

- **3M 1826 surgical mask** (3-ply nonwoven material with nose wire and ear loops) **filters out approximately 75% of the particles** that a N95 respirator filtered.
- **Medical/Dental masks** with a nose wire and ear loops and specification sheet indicating 3-ply polypropylene that was purchased online from Staples (No product number available) **filtered approximately 60% of the particles** that the N95 respirator filtered.
- **Cloth masks** (2-ply cotton and using various fabrics that were loose fitting) **filtered approximately 30-60% of the particles** that the N95 respirator filtered.

Initially, public health officials, politicians and many in the media reported that wearing masks does not help the person wearing the mask. This does not appear to be the case or backed up by science. It is not believed that this was done intentionally to mislead the public. It is only mentioned because this may explain some of the confusion in the public as to whether masks are necessary or why some people do not wear face masks. It may also explain the distrust the public may have of public officials, health officials and the media. To be clear, **wearing a mask DOES help protect the person wearing the mask** to varying degrees. Wear a mask, it can help others and you.

Masks are especially important when you cannot maintain proper social distance between you and others. Masks should be used in conjunction with distancing not as a replacement for it. It is not a choice of one or the other. Also, masks help provide some protection for you and others in the event someone unexpectedly coughs or sneezes (wearer or others).

# How Much Do Cloth Masks Protect Us?

A cloth mask does not provide the same level of protection as a respirator or a surgical mask but much better than no mask at all. A cloth mask made of more than one layer of fabric is typically more protective than a single fabric layer cloth mask i.e. a scarf or bandana. According to the Northwestern University article cited in the surgical mask section, cloth masks provide some protection to the wearer of about 30-60%. It can **filter of out 30-60% of the particles that a N95 respirator filters out**.

# Who Does the Mask Protect?

The **mask protects the person wearing the mask and those around the person who is wearing the mask.** As mentioned before, this was not the message that many heard from public health officials, politicians, and members of the media early in the pandemic. It is believed that was because they did not know any better. At the time, this was the belief of many, even experts. Although one might say that common sense would lead one to believe that masks could provide some protection. Some allege that it was known that masks helped but was intentionally not communicated in order to "save" masks for healthcare workers and first responders. While one may never know what others intended, this seems unlikely. When it was later recommended to wear a mask to help others, public health officials, public leaders and media outlets still communicated that it was not to protect the wearer. If the public followed the recommendations or guidelines, they would have used masks to protect others and there would be less available for healthcare workers or first responders. It seems more likely that they simply did not know any better at the time.

# How to Properly Put On a Face Mask?[xv]

- **Wash your hands** before touching the mask or attempting to put on mask
- Hold the mask in front of your face at correct level (so at level of the bridge of your nose and below mouth, or under your chin)
- Loop the ear loops behind your ears (if loops), or tie the tie behind your head if mask requires tying
- Try to **fit the mask as snug as possible** to your face on the sides, under your chin and at the bridge of your nose. If there is a "nose wire," mold it to the shape of your nasal bridge (bony part of the front of your nose). The better the fit i.e. the snugger the mask, the more likely it will better protect you and others.
- **Be sure you can breathe comfortably** while wearing the mask.

- If you cannot breathe comfortably, you may want to avoid going out or choose a different mask. A different mask i.e. thinner, may be easier to breathe through but may not offer as much protection. But **a thinner mask is better than no mask at all.** One must balance the risks and benefits, use good judgment and common sense.
- **Those unable to breathe easily with a mask, under 2 years of age, incapacitated or unable to remove their mask without help should not wear a mask.**

# What You Don't Know About Cloth Face Masks?

- You should **wash your hands before you put on your mask**, if you should touch your mask, and after you remove your mask. Your mask may have virus on it that was filtered. If you touch it, you can contaminate your fingers and then infect yourself or others.
    - Many people remove their mask to eat but do not wash their hands. **I.e. if on an airplane** and you are going to eat or drink something, you should wash your hands or use an effective hand sanitizer before you eat or drink, otherwise **it is possible to contaminate your fingers when you remove the mask and then inoculate yourself with the virus** (if it is on your mask) when feed yourself and touch your mouth or rub your eyes.
- After you remove your mask, you should place it in a plastic bag to prevent it from contaminating other things and inadvertently infecting yourself or others.

- The CDC recommends that cloth face covers **should be washed after each use.**[xvi]
- You can wash it along with your clothes using the **washing machine** and laundry detergent appropriate for the material of the face covering (mask)[xvii]
- You can **wash by hand** using a bleach solution ○ 5 tablespoons (1/3rd cup) household bleach per gallon room temperature water or
    - 4 teaspoons household bleach per quart room temperature water[xviii]
- After washing, you can dry in the clothes dryer on the highest temperature setting until dry or lay flat and allow it to air dry. Placing it in direct sunlight is suggested.[xix] It may help dry and the UV light may kill virus.
- Cloth masks provide the wearer some protection in addition to those around the wearer.
- **Cloth masks** do not serve as a substitute for other preventative measures, they **are to be used in addition to other measures** i.e. social distancing, hand washing, disinfecting surfaces, etc.

# Why Don't Some People Wear Face Masks?

- **Lack of information or understanding**. Most people mean well but many may not understand that it is helpful to others or themselves. They may assume (even if incorrectly) that if they feel good, they cannot have or spread the SARS-CoV-2.

- **Inconvenience**. Some may not want to wear it because it is inconvenient to do so or may forget to bring it along.
- **Incorrect information**. Some may be misinformed that it is not necessary when it is or that it does not help when it does. In fairness to these people, this misinformation was spread by public leaders, politicians, and the media. If someone hears the wrong information, particularly from what should be a credible source i.e. CDC, public health official, they may follow it. Not everyone constantly looks for current updates. These people may be doing what they were "told" to do and not be aware that the public leaders and health officials changed their recommendations or guidelines.
- **Distrust** of those recommending the masks. Based on the current social and political environment and conflicting reports and recommendation, often by the same public officials, health officials and media personalities, some may become skeptical of, or not trust, those sources.
- Some may **not have access to masks**. Not everyone can get a respirator or surgical mask. Typically, these are not recommended for the public. Some may not be able to afford cloth masks or enough of them to always have a clean one ready to use. Recall the CDC recommends washing after each use.

# Why is Good Hygiene so Important with COVID-19?

- Good hygiene is always important to help prevent the spread of "germs" or infection.

- Good hygiene becomes even more useful and important the more contagious a virus is. It appears that the SARSCoV-2 is easily spread. That is part of the reason there are so many cases. **Good hygiene practices are very important to control the spread of SARS-CoV-2 to yourself and others.**

# What is Meant by Good Hygiene?

- **Frequent Hand Washing**
    - **Wash your hands** using soap and water for a **minimum of 20-30 seconds.** The longer you wash them, the more likely you will remove any contamination i.e. virus containing material. Be sure to wash all parts of your hands and not forget about individual fingers, between your fingers, the back of your hands, etc.
    - **Wash Hands BEFORE:** eating or preparing food, touching eyes, nose, or mouth, putting on face mask, caring for others i.e. care provider
    - **Wash Hands AFTER:** taking off mask, touching, or adjusting, mask while wearing it, after eating or touching your eyes, nose, mouth, sneezing, coughing, blowing, or wiping nose, touching or caring for others i.e. handshake, fist bump, care provider etc.
    - Just because you touch something contaminated does not mean you will get infected. The key is to wash your hands often enough and after touching things that may contaminate your hands before you possibly inoculate yourself by touching your eyes, nose or mouth with your hand or fingers that may

have touched something that had the virus on it.

- **Wash Clothing**
  Remove your clothing when you return home if there is a chance they may have been exposed to contamination. Place them in the washer to be washed using regular laundry detergent. For extra safety, use regular laundry bleach unless contraindicated. There is a chance, although not too high, that your clothes can get contaminated, you touch them and then inoculate yourself or others.
- **Disinfect**
  - **Clean surfaces.** Surfaces can get contaminated with SARS-CoV-2 from people touching it with their hand, fingers or other items that may have virus on them OR from droplets that are breathed out and settle on surfaces. This is especially true if someone coughs or sneezes. If you disinfect these surfaces, you reduce the chance of contamination and the likelihood of you, or others, becoming inoculated or infected. Depending on the temperature and other factors, **droplets containing the virus can last for hours or days on various surfaces**. This is one of the reasons why disinfecting and hand washing are so important.[xx]
  - **Use effective disinfectants.** Lysol, Clorox, etc. have disinfectants that kill viruses. In addition, one can use alcohol containing **products containing a minimum of 60% alcohol**. Alcohol used as an "antiseptic" that you can buy at your local pharmacy is typically about 70% or 90%

isopropyl alcohol and should be effective if not diluted.

- **Be sure to use appropriate disinfectants.** Make sure to use certain disinfectants as recommended. For example, you should not use Lysol or Clorox to disinfect your skin but it is perfectly appropriate to use on a countertop, etc.

# Things to Know about Hand Sanitizer

- The name brand is not as important as the contents. In order to be effective, the CDC recommends hand sanitizer should contain an **alcohol concentration of 60% or more.**
- **Make sure the alcohol is not methanol. Methanol, also known as "wood" or "methyl" alcohol" is harmful to people**. As of the time of this writing July 23, 2020, the FDA lists 77 hand sanitizers that contain methanol and should not be used. Go to the FDA website for more information and the most up to date listing. https://www.fda.gov/drugs/drug-safety-andavailability/fda-updates-hand-sanitizers-methanol
- **If a hand sanitizer claims to be approved by the FDA, be suspicious and BEWARE! "There are no hand sanitizers approved by FDA.**[xxi]
- Hand washing is preferred but hand sanitizer is a reasonable alternative when hand washing is not available.

- Some have raised the concern that widespread use of hand sanitizer could lead to unrelated problems such as certain organisms becoming more prevalent or not susceptible to certain antibiotics or treatments. I.e. can breed a new generation of bacteria that are not presently a problem. This is one reason why hand washing is preferred.

- Given the risks and benefits of using hand sanitizer during the COVID-19 pandemic, if hand washing is not an option, I would personally use hand sanitizer for the benefit of killing any virus that may be on my hands and assume the risk of using it.

# *Risk Factors of Getting Infected

If we assume everyone is a carrier, there are many factors that influence the risks for being exposed including but not limited to:

- **Distance** between you and the other person. The greater the distance, the less likely the chance of spread.
- **Time or duration of exposure.** The shorter the time one is exposed to others, the lower the risk.
- **Face mask**. If everyone is wearing a mask (cloth or surgical) or a respirator, the less the risk.
- **"Illness.**" If someone is coughing or sneezing, the more risk there is for spread.
- **Location**. If you are outdoors, the less likely to spread than indoors.
- **Room Size**. The larger the room, the less likely the risk.
- **Ventilation/circulation of room**. The better the ventilation and filtration of the room, the less likely the

risk. This is not to be confused with a fan that could blow droplets farther between an infected and uninfected person and could potentially increase the risk.

- **Viral load**. The more virus that a person "sheds" or produces, the more likely it seems he or she would spread and thus the higher the risk. Unfortunately, we cannot know this from looking at someone. It is assumed that someone showing signs or symptoms spreads more virus and thus would be a greater risk to spread to others, but this is not known for sure.
- **Immunity.** Depending on the immunity of an individual, you may be more, or less, likely to become infected from an exposure. We still do not know for sure what characteristics and qualities makes one more, or less, likely to become infected. We do know certain underlying medical conditions can make the illness worse if one becomes infected with SARS-CoV-2 but it is unclear if these same conditions makes one more likely to become infected. Regardless of the case, everyone needs to take precautions, especially those with underlying conditions or over 55 years of age. Children are much less likely to have significant disease (COVID19) from SARS-CoV-2 but we do not know for certain how likely they are to become infected and be an asymptomatic carrier. There is also some question as to whether children can spread it as easily as adults. This is part of the reason that many believe that it is relatively safe and better (taking into consideration all the pros and cons) to open the schools and allow children to attend in person. Many other countries have re-opened schools and do not seem to be having significant consequences from doing so.

- **Habits.** The less one touches one's face, rubs their eyes or nose, or puts their fingers in their mouth, the less the risk. The more one touches things i.e. surfaces, the greater the risk.
- **Hygiene.** The more one washes his or her hands, the less likely the risk. This is especially true if make sure one washes hands **before** touching eyes, nose, mouth, brushing teeth, eating, preparing food, putting on one's mask, touching another individual i.e. to provide care to them or **after** coughing or sneezing, touching your eyes, nose or mouth, touching another individual, commonly touched items i.e. door knobs, handles, countertops, table tops, remote controls, toilet seats, laptops, keypads, telephones, or anything that could be contaminated including your mask after wearing it. Be sure to use soap and water for **at least 20-30 seconds** and to make sure you do not re-contaminate your hands by touching the sink handles i.e. water control, or door knobs that may not be free of virus or other contagious organism. If you are unable to wash your hands, use a hand sanitizer that kills the virus. For example, a sanitizer that contains a minimum of 60% alcohol.[xxii] Make sure to NOT use hand sanitizer that contains methanol. This is poisonous and some hand sanitizers have been recalled because they contain it.
- **Disinfect.** The more you clean, the less the risk of contracting SARS-CoV-2. The more frequently one cleans, especially commonly touched items, the less the risk.

*This assumes all other things being equal. Any one factor can decrease the risks. Combining multiple factors can decrease one's risk even more.

I hope this booklet provides you with useful information. As mentioned, it came about after the suggestion and recommendation of one of my loyal readers and is intended to serve as a quick reference and to explain some of the history, logic, and reasoning behind the guidance during the Coronavirus Pandemic a.k.a. COVID-19 or SARS-CoV-2.

If you want more information about the history, naming or origin of SARS-CoV-2, the different types and accuracy of testing, or the number of deaths attributed to COVID-19 and how these numbers are derived and the many reasons why these numbers may be wrong, etc., you can find this information in *COVID-19 Lockdown: Unreported Truths & Perspective.* Just search by the title name, my first and last name (Etch Shaheen) or Dr. Etch.

Thank you for your support and helping prevent the spread of SARS-CoV-2 and decreasing the chances of COVID-19 harming you or others. Please tell others what you have learned and help them be able to make better informed decisions and choices. Together, we can and will beat this virus!

Stay safe and well,
Dr. Etch

[i] Marfin, Catherine. July 23, 2020. <
https://medicalxpress.com/news/2020-05-feet-social-distancing.html>
[ii] Foley, Katherine Ellen. July 23, 2020. <
https://qz.com/1831100/where-does-the-six-feet-social-
distancingguideline-come-from/> [iii] CDC. July 23, 2020. <
https://www.cdc.gov/coronavirus/2019ncov/travelers/after-travel-
precautions.html> [iv] Marfin, Catherine. July 23, 2020. <
https://medicalxpress.com/news/2020-05-feet-social-distancing.html>
[v] Morris, David Z. July 23, 2020.
<https://fortune.com/2020/05/20/why-6-feet-apart-social-
distancingcoronavirus-cdc-six-foot-rule-spread-covid-19-liquid-
droplets/> [vi] Foley, Katherine Ellen. July 23, 2020. <
https://qz.com/1831100/where-does-the-six-feet-social-
distancingguideline-come-from/> [vii] Foley, Katherine Ellen. July 23,
2020. < https://qz.com/1831100/where-does-the-six-feet-social-
distancingguideline-come-from/>
[viii] Letzter, Rafi. July 20, 2020. <
https://www.livescience.com/coronavirus-six-feet-enough-
socialdistancing.html> [ix] Morris, David Z. July 23, 2020.
<https://fortune.com/2020/05/20/why-6-feet-apart-social-
distancingcoronavirus-cdc-six-foot-rule-spread-covid-19-liquid-
droplets/> [x] Foley, Katherine Ellen. July 23, 2020. <
https://qz.com/1831100/where-does-the-six-feet-social-
distancingguideline-come-from/> [xi] Bourouiba, PhD, Lydia. July 23,
2020.
<https://jamanetwork.com/journals/jama/fullarticle/2763852
[xii] Marfin, Catherine. July 23, 2020. <
https://medicalxpress.com/news/2020-05-feet-social-distancing.html>
[xiii] Shaheen MD, Etch. *COVID-19 Lockdown: Unreported Truths & Perspective.* Knowledge is Power Publishing 2020. [xiv] Mueller, Amy,
Fernandez, Loretta. July 23, 2020. <
https://www.medrxiv.org/content/10.1101/2020.04.17.20069567v2.f
ull.pdf> [xv] CDC. July 23, 2020. <
https://www.cdc.gov/coronavirus/2019ncov/prevent-getting-
sick/how-to-wear-cloth-face-coverings.html> [xvi] CDC. July 23, 2020. <

https://www.cdc.gov/coronavirus/2019ncov/prevent-getting-sick/how-to-wash-cloth-face-coverings.html>

[xvii] CDC. July 23, 2020. < https://www.cdc.gov/coronavirus/2019ncov/prevent-getting-sick/how-to-wash-cloth-face-coverings.html> [xviii] CDC. July 23, 2020. < https://www.cdc.gov/coronavirus/2019ncov/prevent-getting-sick/how-to-wash-cloth-face-coverings.html> [xix] CDC. July 23, 2020. < https://www.cdc.gov/coronavirus/2019ncov/prevent-getting-sick/how-to-wash-cloth-face-coverings.html> [xx] Foley, Katherine Ellen. July 23, 2020. < https://qz.com/1831100/where-does-the-six-feet-social-distancingguideline-come-from/> [xxi] FDA. July 23, 2020. < https://www.fda.gov/drugs/drug-safety-andavailability/fda-updates-hand-sanitizers-methanol> [xxii] CDC. July 23, 2020. <https://www.cdc.gov/handwashing/whenhow-handwashing.html>

Made in United States
North Haven, CT
04 November 2026

21771748R00018